Contents

Simple Machines Work

Do you know how people make work easier? Check out page 5.

Do you know what a force is? Find out on page 5.

Read these words then check out pages 30 and 31.

fulcrum	inclined plane	lever	pulley
FUHL-kruhm	in-KLYND PLAYN	LEE-vuh	PUHL-ee

Do you know how a skateboard works? Read page 7.

Do you know how workers built the Great Pyramid in Ancient Egypt? Go to page 12.

Do you know how workers watered the Hanging Gardens of Babylon? Check out page 20.

screw	wedge	wheel and axle	work
SKROO	WEJ	WEEL and AK-suhl	WERK

HARD at WORK

Written by Nicole Ward

4

You might not think so, but every day you use force to move, stop, or change an object. You do **work**.

Think about rollerblading. Your feet push against the ground to make a force that moves the wheels. That's work!

When you rollerblade, you do work!

You do work every time you use force
to do something.
You use a push force or a pull force.
The push or pull force makes things move,
stop, or change.
You can make the work easier.
You can use simple machines
to help you do the job.
Read on to learn about the six simple machines.

Bikes are made from
a lot of simple machines.

Ramp, or Inclined Plane

A simple machine that helps you lift a load
is an **inclined plane**.
An inclined plane has a high end and a low end.
An inclined plane helps you lift
a load with less force.
This van has a ramp.
You can push the wheelchair
up the ramp.
You do not have to use much force
to lift the load up the slope
and into the van.

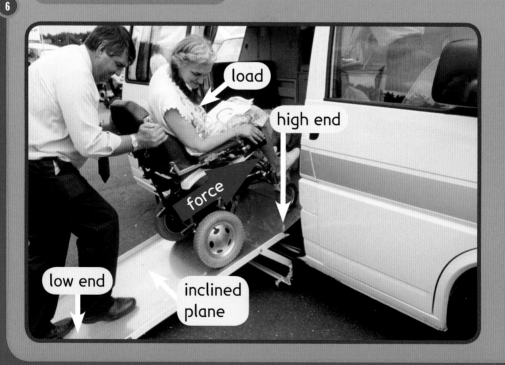

load

high end

force

low end

inclined
plane

Wheel and Axle

A simple machine that helps you move a load is a **wheel and axle**.
Your skateboard uses a **wheel and axle**.
The wheel turns on a pole.
The pole is called the axle.
The wheel and axle are joined.
They turn around together.
You use a push force with your foot to move the wheels.
The skateboard moves the load of the rider easily.

Your skateboard is a machine with a wheel and axle.

axle

wheels

Screw

A simple machine that joins things
is a **screw**.
A screw has a sloped ridge around a pole.
The ridge is called a thread.

You turn the screw.
The thread cuts a path into the things it joins.
The screw holds the things together.

Screws hold the cover
on this box.

force

screw

screwdriver

Wedge

A simple machine that cuts
or splits things is a **wedge**.
A wedge is a block with sloping sides.
It may have a sharp edge or point at the end.
A wedge can help change
the direction of a force.
A knife is a wedge.
You can't slice tomatoes with your hands.
You take a knife.
You push down.
The knife helps you do the job.

9

When you push
this wedge down,
you force the slices
of tomato apart.

force

movement

sloping
sides

Lever

A simple machine that helps you lift a load is a **lever**.
A lever is a straight bar.
It turns on a fixed point, or **fulcrum**.
If you use a lever, you need less force to lift a load.

A seesaw is a lever.
It moves on a fixed point called a fulcrum.

load

You use force on this end.

force

fulcrum

movement

bar

The load on this end moves up.

Pulley

A simple machine that helps you lift a load is a **pulley**.
A pulley has a wheel and a rope.
The rope hangs over the wheel.
You tie a load to one end of the rope.
You pull down on the other end.
You lift the weight up.
If you use a pulley,
you need less force to move a load.
A tow truck uses a pulley.
A car is heavy to lift without a pulley.

You can use a pulley to lift a heavy load.

wheel

force

rope

movement

load

Read on to find out how people made an amazing building →

How People Built the Great Pyramid

Written by Nicole Ward

Can you imagine a building
made with 2 million big blocks of stone?
Each block weighs over 2 tonnes.
The building is huge.
It was hard to build.
Thousands of workers did the job.
This building is the Great Pyramid in Egypt.

Ancient Egyptians built the Great Pyramid 4,600 years ago. It is 137 metres tall.

People think workers used
simple machines to build the pyramid.
They think the workers used wedges.
They think the workers used levers.
They think they used inclined planes.

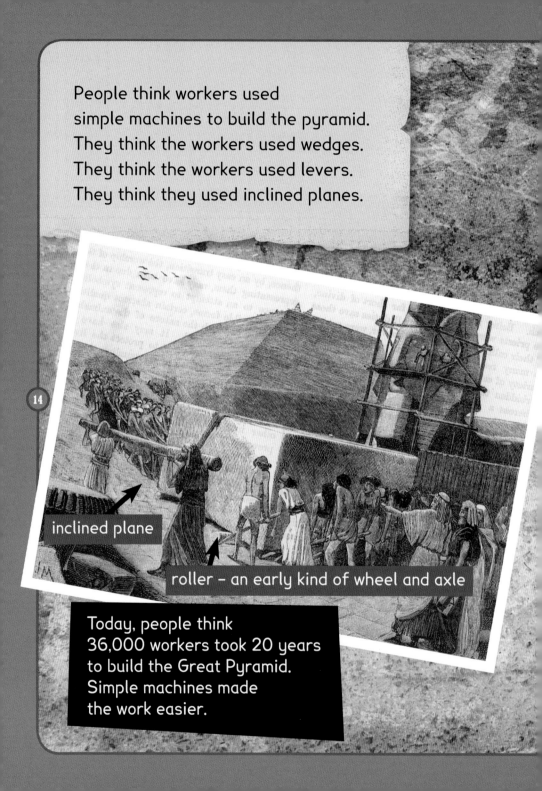

inclined plane

roller – an early kind of wheel and axle

Today, people think
36,000 workers took 20 years
to build the Great Pyramid.
Simple machines made
the work easier.

Stonemasons used wedges.
They cut big bits of limestone
from the limestone mine.
They marked where to split the stone.
They hit a wooden wedge called a chisel
into the mark.
They used a wooden hammer to hit it.
The hammer had a big head.
It was called a mallet.
The stone split.
That way, stonemasons could cut big blocks
of stone into smaller blocks.

15

Using a wedge
to split stone

wedge

The blocks were down on the ground.
The builders were up on the building.
Workers built a ramp, or inclined plane,
up to the builders.
They built the ramp from dirt and stones.
They put pieces of wood across the ramp.
They poured water on the ramp.
This made it smooth and slippery.

Workers used
an inclined plane to help
shift stone blocks.

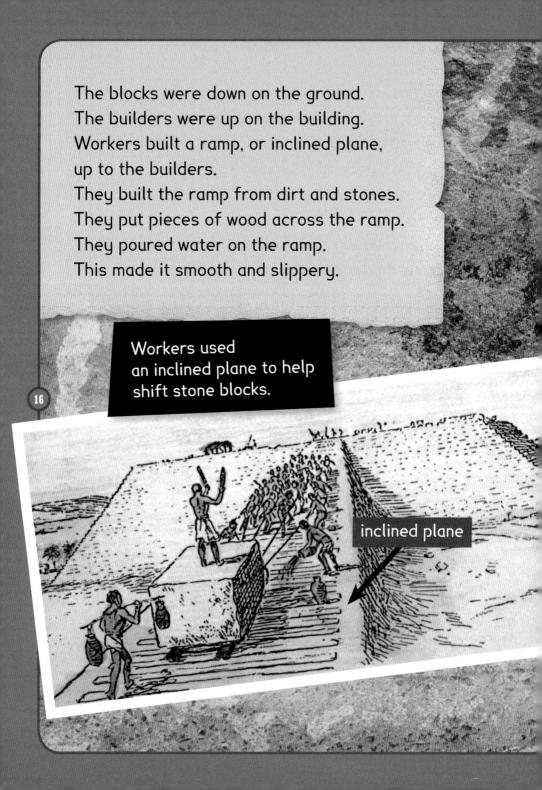

inclined plane

Workers used ropes and levers
to put a stone block on a wooden sledge.
They pulled the sledge up the ramp.
They used levers to put the stone block
in the right place.
Workers made the ramp higher
to move the blocks of stone higher.
The building grew taller.

Simple machines helped people
build the Great Pyramid.

lever

Workers used
wooden levers
to help lift
stone blocks.

Read on to find out about the Hanging Gardens. →

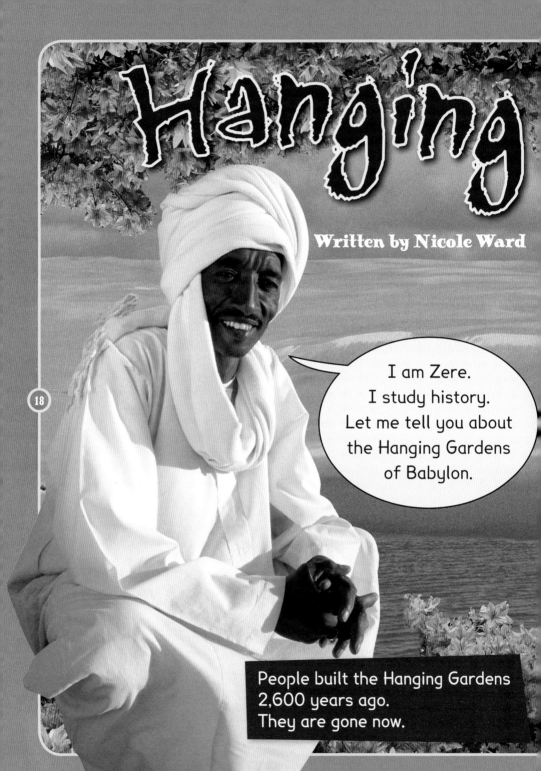

Hanging

Written by Nicole Ward

I am Zere.
I study history.
Let me tell you about
the Hanging Gardens
of Babylon.

People built the Hanging Gardens
2,600 years ago.
They are gone now.

Gardens

The Hanging Gardens grew on a high building.
The building was in the desert.
The gardens grew on many floors, or terraces.
Big trees and flowers grew in the gardens.
Look at the drawing of the gardens.

19

The Hanging Gardens needed water.
The gardeners used a simple machine
to get water to the plants.
They used a chain pump.
This is a drawing I made of a chain pump.

A Chain Pump to Lift Water

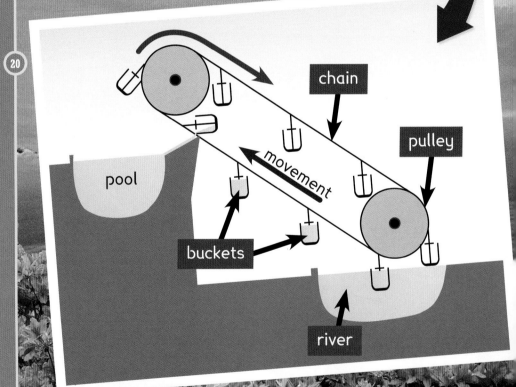

chain

pulley

pool

movement

buckets

river

The chain pump used a pulley.
It moved many buckets.
The buckets took water from the river.
The pulley pulled the buckets to the top of the gardens.
The buckets tipped the water into a pool.
Water flowed down channels to the terraces.
The channels took water to all the plants.

Simple machines helped people
grow the Hanging Gardens of Babylon.

Read on to find out about a strange gift. →

Gift from the Gods

Written by Nicole Ward

Illustrated by Des Young

It had been a hard day's work at the Great Pyramid. Ako, Rasui, and Urshe had been pulling stone blocks to the Great Pyramid. Now they were resting.

Ow! Did you hear that? That was my back! My back hurts!

CRACK!

Your back always hurts. But we are building the place where our king wants to be buried.

Yes, but I am sure there must be an easier way!

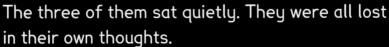

The three of them sat quietly. They were all lost in their own thoughts.

So, what if...what if we asked the gods for a gift?

What are you talking about?

Well, if I could ask the gods for a gift, I would ask them for a big hippopotamus.

Exactly how would a big hippopotamus help us?

We could tie wooden planks to the hippopotamus's feet. Then it would push the sand down as it walked.

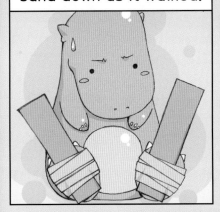

Now who is the dreamer, Ako?

25

What Ako, Rasui, and Urshe did not know was that what they talked about came true. Today, this machine is called a bulldozer!

Read on to learn about simple machines working together. →

Multimedia Information

www.readingwinners.com.au

FAQS- - - - - - - - - -

Q What is a machine that uses more than one simple machine called?

A A compound machine.
A digger is a compound machine.
It is made up of many simple machines.
It uses wheels and axles, levers, screws, and pulleys.

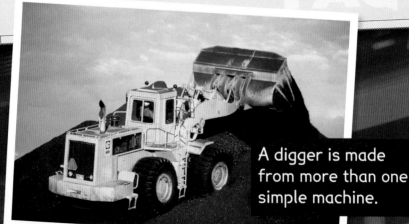

A digger is made from more than one simple machine.

Message 1

Hi Tom, here's how 2 fix ur bike.
Put chain back on pedal teeth.
Chain connects pedals & rear axle.
Ur feet push pedals 2 move chain.
Chain turns axle.
Axle turns wheel.
Bike moves.
Difficult 2 hook chain back.
Hook all chain links u can.
Find a lever.
Screwdriver in ur toolkit
will work.
Hook scrwdr under chain
& over teeth.
Lift last few links
onto teeth with scrwdr.
Bike should go now.
Look at pic I dwnlded. *bike.jpg*

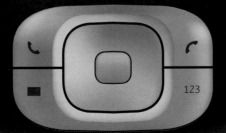

BACK   REPLY

Turn the page to check what you have learned. →

Quick 8 Quiz

1. What does an inclined plane have?

2. How does the wheel and axle work?

3. Why do screws have a thread?

4. Name one kind of wedge.

5. What is a fulcrum?

6. Name one use for a pulley.

7. Name two simple machines that people may have used to build the pyramids.

8. How did people move water to the Hanging Gardens?

Turn to page 32 for clues. →

Learn More

Choose Your Topic
Choose one simple machine from the book.

Research Your Topic
Find out all the ways you can use it.

Write Your Article
You may need to make notes first.
You may need to draw diagrams.
Get your facts in order.
Use subheadings to help you do this.
Write a draft.
Check your spelling.
Check your punctuation.

Present Your Topic
Share your work with other members of your group.

fulcrum – the fixed point on which a lever turns

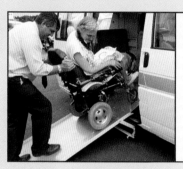

inclined plane – a sloping surface or ramp that has a high end and a low end

lever – a bar that turns on a fixed point

pulley – a wheel with a rope that helps lift a load

screw – a pole with a thread around it used to join things together

wedge – a tool with sloping sides used to split or cut things

wheel and axle – a wheel that turns on a pole

work – the result of force moving, stopping, or changing an object

Index

Clues to the
Quick 8 Quiz

1. Go to page 6.
2. Go to page 7.
3. Go to page 8.
4. Go to page 9.
5. Go to page 10.
6. Go to page 11.
7. Go to page 14.
8. Go to page 20.